Institute of Terrestrial Ecology
Natural Environment Research Council

The Culture and Use of Free-living Protozoa in Teaching

Frederick Page
Institute of Terrestrial Ecology
Culture Centre of Algae & Protozoa
Cambridge
England

Printed in Great Britain by
The Lavenham Press Limited, Lavenham
©Copyright 1981

Published in 1981 by
Institute of Terrestrial Ecology
68, Hills Road, Cambridge, CB2 1LA

ISBN 0 904282 52 X

Cover

At the top: Pupils of Manor School, Cambridge, studying
protozoa in the biology laboratory. Photograph P. G. Ainsworth.
Centre right: *Vorticella microstoma*; bottom right: *Stentor coeruleus;* bottom
left: *Amoeba proteus*. Photographs F. C. Page.

Acknowledgement

The Institute is indebted to the Headmaster and Messrs. B. Hunter & G.
Woodcock of Manor School, Cambridge, for generously providing facilities
for making the top cover photograph.

The **Institute of Terrestrial Ecology (ITE)** was established in 1973, from the
former Nature Conservancy's research stations and staff, joined
later by the Institute of Tree Biology and the Culture Centre of Algae and
Protozoa. ITE contributes to and draws upon the collective knowledge
of the fourteen sister institutes which make up the *Natural
Environment Research Council,* spanning all the environmental sciences.
The Institute studies the factors determining the structure, composition
and processes of land and freshwater systems, and of individual plant
and animal species. It is developing a sounder scientific basis for
predicting and modelling environmental trends arising from natural or
man-made change. The results of this research are available to those
responsible for the protection, management and wise use of our natural
resources.
Nearly half of ITE's work is research commissioned by customers, such
as the Nature Conservancy Council who require information for wildlife
conservation, the Department of Energy, the Department of the
Environment, and the EEC. The remainder is fundamental research
supported by NERC.
ITE's expertise is widely used by international organisations in overseas
projects and programmes of research.

Dr. F. C. Page
Institute of Terrestrial
    Ecology
Culture Centre of Algae
and Protozoa
36, Storey's Way
Cambridge
CB3 0DT
England

Foreword

The primary purpose of this little book is to make possible the maintenance of small collections of protozoa for educational use, particularly in schools and colleges whose circumstances make it impractical to order cultures as needed. To encourage such use of protozoa, I have tried to indicate without great detail how they can serve for illustrating biological principles and for experimental material.

Since this publication is intended especially to encourage the wider use of free-living protozoa for teaching in the Southern Hemisphere, special attention has been given to conditions in warmer countries, such as temperature and availability of components for media. However, I have also drawn on my own experience of teaching in the Northern Hemisphere and hope that this book will be useful in diverse parts of the world.

Many of the methods have been developed at the Culture Centre of Algae and Protozoa; others are adapted from the literature but have been used at the CCAP. I wish to thank the Plant Breeding Institute (Trumpington, Cambridge) and their information officer Mrs Joan Green for information and samples of grains. I especially thank Mr J. P. Cann for sharing his experience with the strains maintained at the CCAP, and for reading the manuscript.

Contents

Introduction

Without the use of living material, it is indeed difficult to make the teaching of biology come to life. However, there are often obstacles to the use of animal material, and bacteria and plants, useful though they are, do not present a complete perspective. Free-living protozoa offer an ideal answer to the need for animal-like material. They can be maintained with little special equipment, in restricted space, with economy of time. They permit direct observation of the structure of living cells. They can be used to demonstrate and experiment with fundamental biological processes such as locomotion, feeding and digestion, exchange of substances across membranes, osmoregulation, and reaction to stimuli. With protozoa, cell division, sexual processes, and even simple morphogenetic processes are readily observable and often easily accessible to manipulation. Protozoa can be cultured in numbers large enough for experiments in physiology, biochemistry, and ecology. To those who object to the trivial use of mammals and other higher animals, they provide especially suitable material for use with beginning students.

Not least of all, to those fascinated by the diversity of life forms produced by organic evolution, free-living protozoa offer a wide range. As a group they are ubiquitous in fresh water, in the sea including some at great depths, and in all soils except perhaps the harshest deserts. One aim of this publication is to encourage teachers to make their students aware of the many protozoa which are not parasitic but rather, as primary and secondary consumers, are parts of the same trophic systems as the larger metazoans with which they are more familiar. Free-living protozoa are part of that invisible world with which we are in daily, intimate contact, a fact which the suggestions for isolation from natural sources are intended to emphasise.

In short, this booklet contains detailed instructions for maintaining a permanent basic collection of free-living protozoa. It includes simple but adequate methods requiring minimal equipment and supplies as well as standard methods which may require somewhat more equipment and a few more components for media but will in some cases be more suitable for experimental purposes. Although this booklet is not meant as a manual of experiments and demonstrations, some uses for the organisms will be suggested. Other uses will be found in syllabi or drawn from the teacher's experience. Those who wish to go more deeply into protozoology will find guideposts in the References. The brief descriptions and illustrations are not a substitute for the more detailed information found in textbooks but merely an indication of the nature of organisms which may not be familiar to all readers.

Given a decision to use protozoa in teaching and assuming access to essential equipment and supplies, the principal remaining practical problem is obtaining seed cultures to start the local collection. The complete answer is not available as this is written. One possibility is to obtain cultures for some central point, such as a college or a large school, and distribute subcultures to others in the area. Certain of the protozoa described (for example, *Paramecium caudatum* and *Naegleria gruberi*) are cosmopolitan, common, and easily isolated from nature. In regions where the protozoa suggested are not indigenous, readers may be able to isolate appropriate indigenous substitutes for some.

Since the biogeography of protozoa has been investigated mainly in limited parts of the Northern Hemisphere, it seems scientifically desirable to delay as long as possible any further obscuring of the picture by dispersal of organisms. Although protozoa are believed to be more or less cosmopolitan and undoubtedly have been dispersed widely by natural events and human activities, there are probably differences in the distribution of species. We suggest therefore, that readers may wish to sterilise glassware, including microscope slides, which has contained organisms not indigenous to their region before discarding the contents or washing the glassware.

Communications from readers are welcome: problems encountered, suggestions for improvements, findings of possible substitute species indigenous to the reader's region, and information on the distribution of protozoa.

Principles and practices

Efforts to culture free-living protozoa are aimed at providing a micro-habitat in which they will reach an adequate population level and remain there for a reasonable length of time. Success requires satisfactory conditions with respect to biological, chemical, and physical factors, just as the fate of a population in the natural environment depends on those factors. Such important factors as pH and mineral nutrients should be adequately provided for with the media and methods proposed, but we must consider a few major aspects first.

Biological classification of cultures
Cultures are often classified according to the presence or absence of other organisms accompanying the species which we intend to culture. This method of classification seems less strange when we consider that any other organisms present are likely to become food for the desired species, or prey on it, or compete with it for food, oxygen and other needs, or pollute its environment with carbon dioxide and other waste products.

The ideal culture is often the *axenic* one, that is, with no other species present. The organisms are usually absorbing nutrients in solution, sometimes ingesting inanimate particles. Examples are the methods presented here for *Peranema* and *Tetrahymena* and one method given for *Chilomonas*. Some of the other organisms discussed in this publication have also been cultured axenically, but the methods and media used for them are not practical for present purposes. The ultimate in axenic culture is use of a *defined medium*, one whose chemical composition is known exactly. Defined media are used in determining the nutritional requirements of organisms, that is, those substances which must be supplied because they cannot make them themselves. None of the media suggested in this booklet are defined.

A *monoxenic* culture contains one other species besides the one which we intend to culture, usually the food organisms. An example is *Naegleria gruberi* with *Escherichia coli*. A *polyxenic* culture contains several other species; if their identities are not known it is *agnotoxenic* or *agnotobiotic*. Ciliates are especially likely to be in polyxenic culture, often agnotoxenic. It is not always advisable to try to clean out every tiny flagellate or small ciliate accompanying a larger protozöon, at least not until we are sure that it is not serving as food or helping to clean out excess bacteria. Of course, if one wishes to investigate food preferences of a predator species, one must remove other organisms. Such studies are a practical sort of experimental work for school and college laboratories.

The presence of food organisms complicates matters by making it necessary for us to provide conditions suitable for their adequate but not excessive growth. Sometimes we may add food organisms in about the right quantity without providing for them to multiply. An example is streaking *Escherichia coli* on non-nutrient agar to feed *Naegleria gruberi*. But in the majority of cultures described here, the food organisms and sometimes others accompanying the desired species are transferred along with the latter into fresh medium.

Kinds of media

The media suggested in this booklet are of several principal kinds, which may overlap. I have already mentioned media for axenic cultures. The medium for *Tetrahymena* is a complex organic solution. That for *Peranema* is a soil extract supplemented with milk, itself a complex organic mixture.

Infusions of plant matter are the most common methods of growing free-living protozoa, especially ciliates. Such infusions are often made with hay, dried lettuce leaves, or the commercial cereal grass preparation Cerophyl. The reader may wish to try one or more of those materials. However, we have found infusions of grains very useful. They are certainly the easiest media to prepare, and their components are common materials. The grains most commonly used are polished rice, barley, and wheat. We have tried these and unpolished rice, millet, maize (dried kernels), and lawn grass seed for several species of protozoa. The most surprising observation has been the usefulness of maize, which has not been used for this purpose before, as far as I know. The grains generally provide food for the bacteria on which the protozoa feed. In the case of *Chilomonas* they must provide food directly for the flagellates, which do not feed on the accompanying bacteria. They may provide some nutrients directly to other protozoa or influence the bacterial species present or chemical and physical factors such as the pH.

Soil extract is prepared by heating water with soil, a process which should dissolve some substances and kill the organisms present and must also change some organic constituents chemically. Some of our strains are cultured in *biphasic soil extract* medium, that is, with a layer of soil left at the bottom of the tube or the beaker. At this point the distinction between grain infusions and soil extracts is blurred, because these cultures in biphasic soil extract contain a few grains, usually barley. The reader may wish to experiment with other grains in biphasic soil extract cultures. The grains are usually added before the medium is steamed, but polished rice should not be added until after steaming, since it is likely to disintegrate.

In general, we consider that organisms become numerous more quickly in grain infusion cultures, but biphasic soil extract cultures, particularly larger ones in beakers, remain stable longer after achieving a high level. In the cases of those species for which a biphasic medium is used, some attention as suggested in the instructions for *Paramecium* and *Stentor* is essential. If the population is declining, we may put some organisms into an infusion in a small culture dish and then inoculate a beaker of biphasic soil extract after the organisms have become numerous in the infusion. However, the reader will note in the discussion of several species that infusion cultures may remain abundant for months: *Chilomonas* or *Colpidium* in almost any infusion, or *Stentor* on a maize kernel. It would be possible to maintain a collection including most species discussed in this booklet entirely in infusions, in numbers adequate for teaching and without excessive demands on the teacher's time, given careful attention to the components of the media and to such factors as temperature.

The remaining categories of media described in this booklet are *natural water*, with or without a chelating agent, for organisms which must be fed at fairly frequent intervals, such as *Actinophrys*, and *agar* media, as suggested for *Saccamoeba limax* and *Naegleria gruberi*, though both these organisms can be grown in liquid media if agar, petri dishes, sterilising equipment, and food bacteria are not readily available.

Many small and medium-sized amoebae are grown on agar. The preferred method is to grow them on non-nutrient agar, which is streaked just before the amoebae are added with the food bacteria, which have been grown on nutrient agar slopes. One should not try to save a step by growing the amoebae themselves on nutrient agar, because the bacteria are likely to overgrow them and one will never be able to get microscopical preparations without excessive bacteria. These agar cultures of amoebae should be handled with as much care for aseptic conditions as are bacteria, but one should always transfer amoebae on a block of agar or in a pipette as a cell suspension washed off the agar with about 1 ml of liquid, never with a bacteriological loop. The condition of the amoeba culture (multiplication, migration, encystment) can be seen by turning the petri dish over on the microscope stage and focussing through the agar on to the surface. More details on the culture of amoebae are given by Page (1976).

Temperature
This is one of the principal limiting factors affecting growth and survival of cultures, easily measured but less easily controlled.

All organisms have a range within which they can live and an optimum at which they do best. Some protozoa do well over a fairly wide range of temperatures. However, since many protozoa, like many other organisms including man, tend to live near the upper part of their range and even to have their optimal temperature toward that end, a moderately increased temperature may well be a limiting factor for cultures. Apart from any directly lethal effect of high temperatures, cultures are affected by the influence on metabolic rate. Protozoa which multiply over a wide range of temperatures generally multiply more rapidly toward the upper part of that range, though not very near to the limit. If the rate of any process such as encystment is related to the population growth rate, that process, too, will happen sooner at the upper end of the range. But if the protozoa do not have an encysted state to tide them over until they are put into fresh medium, rapid multiplication will mean rapid exhaustion of food supply and rapid achievement of any other time-dependent unfavourable conditions, such as accumulation of protozoan or bacterial waste products.

The practical lessons from these facts are: (1) The temperature obviously must be kept within a range suitable for survival. (2) The temperature must permit sufficient growth to reach the desired population level. If the culture is to be used immediately, one will wish to reach that level sooner than if the culture is to be kept as a source of inocula for future cultures. (3) One can keep a culture longer if, after it has reached a population level high enough to make extinction unlikely, it is stored at a lower temperature still within its tolerated range but low enough to reduce metabolic rate and population growth significantly.

In adapting our methods for use in diverse situations, I have kept in mind particularly the problems of teachers in warmer countries and tested some of our strains for growth at 20° and 30°C. Some would not tolerate prolonged (that is, a week's) exposure to a constant temperature of 30°. We know from experience that these strains have at times survived (in depleted numbers) day-time temperatures over 30°C; however, in such cases the night-time temperature was lower.

Information on ability to tolerate a constant temperature of 30°C is given with each species, but the different strains may be grouped as follows: (1) maintained permanently at room temperature, with a day-time maximum about 25° and usually less: *Peranema trichophorum, Saccamoeba limax, Actinophrys sol, Colpidium striatum, Paramecium caudatum, Stentor coeruleus,* and *Vorticella microstoma;* (2) maintained at a constant temperature about 18° or 19°C: *Amoeba proteus;* (3) grown for a few days at room temperature, then stored at 13° to 15°C: *Chilomonas paramecium, Tetrahymena;* (4) grown at room temperature, then stored for up to two or three months on an agar plate at room temperature or for a year on an agar slope at 63 to 9°C: *Naegleria gruberi.* In addition, reserve cultures of some of the first group are stored at 13° to 15°C; *Saccamoeba limax* can be stored for months on an agar slope at 9°C; *Tetrahymena* will often survive several months at 4° to 6°C.

There are several possible solutions to the problem of temperature in culture maintenance: (1) If possible, let the cultures grow for several days at a temperature which gives good numbers, then store them at a lower temperature. This method will give the longest-lived cultures and reduce the necessary frequency of subculturing. (2) Devise a method of keeping the cultures permanently at a moderate temperature; they will probably survive a somewhat elevated day-time temperature if there is sufficient cooling at night. (3) Use only strains that tolerate the ordinary environmental temperature. This may limit the variety that can be maintained. However, it may be possible to find local strains of cosmopolitan species that will tolerate the ordinary local temperatures, since strains of a species may vary in temperature range. A certain amount of adaptability may make gradual adaptation of non-indigenous strains possible, but that may not be of much significance here.

Other points
When inoculating fresh medium, transfer as many cells as possible without carrying over great quantities of the old medium; a pipette full is the maximum except for *Amoeba proteus,* where more may be needed.

Cultures should never be kept in direct sunlight. Even those which are feeding on algae or contain zoochlorellae and need light should be kept in the shade. Most protozoan cultures can be kept in a dark cupboard, though of course they will tolerate indirect daylight.

Measures should be taken to retard evaporation and minimise entry of contaminants without excluding oxygen. This means attention to covers and plugs. We close most of our tubes and culture dishes, including petri dishes, with polythene clingfilm around the edge of the cap or cover, partly to retard evaporation but especially to keep out small insects, mites, and air currents bearing fungal spores and bacteria. Biphasic soil extract cultures in beakers are covered with greaseproof paper secured with an elastic band. We recommend the petri dishes described as unvented or singly vented rather than the triply vented ones, which permit more rapid drying out of agar as well as easier access to air currents carrying fungal and bacterial contaminants.

The instructions in this booklet should be regarded as guideposts pointing in the right direction rather than prescriptions which must be followed rigidly. The reader who looks for the underlying principles, adapts the methods to his own conditions, and improvises is likely to be most successful. There is much room for modification. For example, the reader may want to determine the maximum interval which he can safely leave a strain without transferring it to fresh medium. In that case, he might start with the minimum time suggested and gradually lengthen it, carefully observing the condition of the cultures but not waiting until they are on the brink of extinction.

Equipment and supplies

A collection can be maintained with little more than a few jars and bottles, using grain, water, and soil that are available locally. As far as possible, the instructions suggest methods which can be adapted to such conditions. Jars with a capacity of 250 or 300 ml can be substituted for beakers, smaller jars for small culture dishes, and medicine bottles for flasks to store liquids, as long as there is no need to place this glassware on a flame. We do recommend that, if at all possible, media not be heated in metal pots. The list of supplies includes the maximum that would be needed to carry out all procedures suggested. Some of these supplies can be omitted with no important disadvantages. Both supplies and preparation will be considerably simplified if only grain-infusion cultures are maintained, although the preparation of soil extracts requires little additional equipment beyond larger jars and some sort of steamer. An asterisk (*) in the lists below indicates items needed for a collection entirely made up of cultures in grain infusions made with natural or tap water.

Obviously, at least one *microscope* is needed before one can even start to look at protozoa. A maximum total magnification of 400 or 500 diameters is adequate, though an oil immersion objective (approximately × 100) is an advantage with the smallest organisms. It is better to spend the same amount for a better microscope with bright-field optics than to introduce phase contrast for teaching. Some of the illustrations were made with differential interference contrast, which is never used for teaching.

A *balance* (or use of one) will be needed for a collection not made up entirely of grain infusions in natural or tap water. A simple *centrifuge* would be useful for such optional procedures as washing cells but is not of high priority. Instructions for making an inexpensive centrifuge are given by Belcher (1978).

The majority of the organisms discussed here can be maintained without sterilising equipment. Agar media and liquid organic media must, however, be sterile, as must the glassware (petri dishes, tubes, pipettes) used with those cultures. Ordinarily steam under pressure above atmospheric pressure is used for sterilisation. A pressure cooker is sometimes used by teachers for small quantities. An autoclave may be available at a near-by health facility. To sterilise glassware which has contained protozoa (see page 6), a chemical agent such as sodium hypochlorite can be used if heat sterilisation is not possible.

Obviously some means of *temperature control*, to keep cultures at temperatures not exceeding 25°C, is highly desirable and will permit the use of a wider range of protozoa, particularly of identified strains originating in the

Northern Hemisphere and useful in connection with many text-books and syllabi. The ideal equipment would be a refrigerated incubator, but such equipment is ordinarily not available in schools or colleges anywhere, and other methods of providing a temperature of 18° to 25°C can be devised. If running water is available, a steady stream passing through a coiled tube inside an insulated box could be used as a cooling mechanism, though such a method will often seem wasteful. This is a good example of a case in which those facing the problem can find a better solution than an outsider can prescribe. Another optional piece of equipment which may be considered where circumstances permit is a refrigerator to store at temperatures of 4° to 9°C some media and some cultures which can be kept much longer that way, such as axenic *Tetrahymena*, bacteria on nutrient agar, and amoebae such as *Naegleria gruberi* on agar.

Subject to the above remarks about substitution, a list of supplies which would make possible all the procedures suggested follows. The reader should consult the sections on specific organisms and media to make his own list from this.

Glassware:
     beakers, 250 ml tall-form for biphasic soil extract, 1 or 2 litre low-form for making media
     *cover glasses
     *covers for culture dishes
     *culture dishes, approximately 50 mm diameter by 30 mm depth, possibly also 90-100 mm diameter by 40-50 mm depth
     Erlenmeyer flasks, 150 ml, 250 ml, 500 ml, and 1 litre
     funnels for filtering
     glassware and tubing to make still for distilled water
     graduated cylinders, 100 ml and 1 litre
     *microscope slides, flat and with concavity or ring for hanging drops
     petri dishes, 90-100 mm in diameter
     *pipettes, Pasteur or similar
     test tubes for liquid media and agar slopes, 150 mm long, 15 mm inner diameter
Miscellaneous items:
     bacteriological loop
     clingfilm (polythene film)
     elastic bands or string to secure greaseproof paper on beakers
     filter paper
     greaseproof paper or equivalent to close beakers containing cultures
     metal caps for tubes (if preferred to cotton wool plugs)
     non-absorbent cotton wool for plugs
     petroleum jelly

*rubber teats for pipettes
soft paper to wrap pipettes for sterilising
*Grains, appropriate ones of the following:
  barley
  lawn grass seed
  maize
  millet
  polished rice
  unpolished rice
  wheat

Complex media components (brand names given only to indicate those which we have used):
  beef extract (Difco)
  non-nutrient agar (Oxoid No. 1 or Difco Bacto-Agar)
  nutrient agar (Oxoid or Difco)
  powdered milk
  proteose peptone (Oxoid or Difco)
  yeast extract (Oxoid or Difco)

Chemicals:
  acetic acid, glacial, CH_3COOH
  calcium carbonate, $CaCO_3$
  calcium chloride, $CaCl_2 \cdot 2H_2O$
  magnesium sulphate, $MgSO_4 \cdot 7H_2O$
  mercuric chloride, $HgCl_2$
  nickel sulphate, $NiSO_4 \cdot 6H_2O$
  potassium chloride, KCl
  potassium dihydrogen orthophosphate, KH_2PO_4
  di-potassium hydrogen orthophosphate, K_2HPO_4
  potassium nitrate, KNO_3
  sodium acetate, CH_3COONa
  sodium chloride, $NaCl$
  sodium EDTA (diaminoethanetetra-acetic acid),
    $[CH_2N(CH_2COOH)CH_2COONa]_2 \cdot 2H_2O$
  di-sodium hydrogen orthophosphate, Na_2IPO_4
  sodium nitrate, $NaNO_3$
  Congo red
  methyl cellulose
  methyl green

Food organisms other than protozoa:
  baker's yeast
  Escherichia coli

Preparation of media

Water
Culture media may be based on natural water collected directly from a stream, lake, pond, or well; tap water from the public water system; or distilled water. The choice depends on circumstances, including the quality of the water available, and the purpose of the medium. Natural water should contain as little contamination from chemicals or animal (including human) waste as possible. It should be filtered very soon after collection and boiled for a short time (or steamed or autoclaved) as soon as possible. If heating produces a precipitate, the water should be filtered again. If possible, water or liquid medium that has been heated should be left standing for a day or so to accumulate dissolved oxygen before being inoculated with organisms. Tap water should contain as little metal as possible and should be left standing uncovered over night or heated to remove chlorine.

Actinosphaerium *medium*
This is simply natural water (prepared as above) to which a chelating agent has been added to prevent precipitation of some dissolved substances. Prepare a stock solution of the chelating agent sodium EDTA (see list of chemicals on page 15) by dissolving 145 mg in 500 ml of distilled water at 90°C. The final medium consists of 1 ml of sodium EDTA stock solution and 999 ml of natural water. In our laboratory, the final solution is aerated by pumping a stream of air through it for two hours before inoculating, but that step can be omitted with *Actinophrys.*

Inorganic salt solutions
PRESCOTT'S AND JAMES'S SOLUTION (for *Amoeba proteus*):
Make up three stock solutions, each with 100 ml of glass-distilled water.

Stock solution A

$CaCl_2 \cdot 2H_2O$	0·433 g
KCl	0·162 g

Stock solution B

K_2HPO_4	0·512 g

Stock solution C

$MgSO_4 \cdot 7H_2O$	0·280 g

Combine 1 ml of each stock solution and 997 ml of distilled water to make 1 litre of the final dilution.
MODIFIED NEFF'S AMOEBA SALINE (for non-nutrient agar):
Make up a separate stock solution of each of the following components by dissolving the quantity given in 100 ml of glass-distilled water.

NaCl	1·20 g
MgSO$_4$·7H$_2$O	0·04 g
CaCl$_2$·2H$_2$O	0·04 g
Na$_2$HPO$_4$	1·42 g
KH$_2$PO$_4$	1·36 g

Prepare the final dilution by adding 10 ml of each stock solution to enough glass-distilled water to make 1 litre.

Grain infusions

Use grain which has not been treated with fungicide or insecticide. Use natural water prepared as described on page 16. Put the water into a glass vessel such as a beaker, flask, or glass saucepan, add the grains, bring to the boil, and boil for the time recommended for each grain. Dispense into culture dishes of the size recommended for each organism (50 mm diameter by 30 mm depth except for *Amoeba proteus*) or approximately equivalent small jars, and let cool to room temperature before inoculating with organisms. Do not put more grains into one dish than the proportions of grains to water given below; filling a 50 mm culture dish or equivalent vessel to a depth of 13 to 15 mm will use 25-30 ml of liquid. One can also boil a number of grains in a relatively small quantity of water, then dispense the grains into water which has already been poured into culture dishes. The quality of the cultures does not seem to differ much whether or not the water is that in which the grain was boiled. Sometimes it is helpful to let the bacteria accompanying the protozoa multiply for two or three days before adding the protozoa themselves, though that is not our usual practice. Sometimes (especially in polished rice infusions) a white mould grows out of the grain, but it is usually harmless to the protozoa and is thought by some workers to be useful in *Amoeba proteus* cultures as a substratum on which the amoebae can move while feeding.

Barley, wheat, unpolished rice: 2 or 3 grains per 25-30 ml water. Boil 5 minutes.

Lawn grass seed: 10 or 12 grains per 25-30 ml water. Boil 5 minutes. I have often boiled a larger quantity of these seeds with the water and then discarded the excess grains after dispensing enough into culture dishes.

Maize: Use dried maize kernels, one per 25-30 ml water. Boil 10 minutes, then tear a hole in the side of each kernel.

Millet: About 5 grains per 25-30 ml of water. Boil only one minute.

Polished rice: 2 or 3 grains per 25-30 ml of water. *Do not boil.* Simply place into water. The grains can be surface-sterilised by passing very quickly through a flame; this will eliminate any mites or undesirable fungi on the surface. If possible, let the rice stand in water for two or three days before inoculating. We use Prescott's and James's solution with polished rice for *Amoeba proteus*.

Soil extract media

Because of the wide variety of soils, I cannot give detailed instructions on choice of soil. In general, neither acid soil nor one with much organic matter or abnormal concentrations of salts is suitable, nor is an excess of sand or clay. We use calcareous garden or agricultural soil. The reader will have to experiment and may find a mixture of two or three different soils suitable. Remove all stones and break or grind the soil into as fine particles as possible.

SOIL EXTRACT WITH SALTS (E + S):

Into a beaker or jar put enough soil mixture and natural or tap water so that the supernatant water occupies approximately four-fifths of the combined depth. Cover loosely and steam for one hour on each of two successive days. Decant or filter the liquid, which is the soil extract. This extract is combined with water and stock solutions of salts to make the E + S liquid, which is useful for growing algae and is the basis of the *Peranema* medium (page 19). E + S liquid consists of

E (soil extract liquid)	10 ml
K_2HPO_4, 0·1 % w/v	2 ml
$MgSO_4 \cdot 7H_2O$, 0·1% w/v	2 ml
KNO_3, 1·0% w/v	2 ml
Glass-distilled water	84 ml

BIPHASIC SOIL EXTRACT:

In tubes: Put a pinch of calcium carbonate at the bottom of the tube, cover with soil to a depth of 1 cm, add a barley grain, and fill to the half way mark or slightly more with natural or tap water. Close the tube with a cap or a non-absorbent cotton wool plug. Steam for 30 minutes on each of three consecutive days, or autoclave once. In 250 ml tall-form beakers or jars such as one-pound jam jars: Calcium carbonate, soil, and water in same proportions as in tubes, with three or four barley grains. We close the beakers with greaseproof paper; if jars are used, leave the lids slightly loose while steaming.

Readers may wish to try other grains if barley is not available. If calcium carbonate is not available, one should nevertheless try the medium, perhaps crumbling a small piece of chalk into the bottom.

FØYN'S ERDSCHREIBER:

Widely used for marine protozoa and algae.

Stock solutions:

    NaNO$_3$ 20 g in 100 ml distilled water

    Na$_2$HPO$_4$ 1·18g in 100 ml distilled water

Final composition:

Filtered seawater	950 ml
Soil extract liquid (E; see above)	50 ml
Each stock solution	1 ml

Liquid media for axenic cultures

CHILOMONAS MEDIUM:

Sodium acetate	1 g
Beef extract	1 g
Distilled water	1 litre

Dissolve solid ingredients, dispense into tubes, close tubes with cotton wool plugs or caps, and sterilise.

PERANEMA MEDIUM:

Add powdered milk, 0·1 per cent w/v (1 g per litre) to E + S liquid (see above), dissolve, dispense into tubes, close tubes, and sterilise.

PROTEOSE PEPTONE/YEAST (for *Tetrahymena*):

Dissolve 20 g of proteose peptone and 2.5 g of yeast extract in 1 litre of distilled water. Dispense into tubes to a depth about two-thirds the length of the tubes, close tubes, and sterilise.

Agar media

After these have been sterilised, whether or not they are poured into petri dishes, they should be stored at room temperature so that any contaminants will grow and become apparent before the agar is inoculated.

NUTRIENT AGAR:

For growing bacteria, such as *Escherichia coli*. Prepare it as instructed by the manufacturer of the nutrient agar powder, using distilled water if possible, otherwise natural or tap water. Pour enough into tubes to fill tubes one-quartor to one-third full, close tubes with cotton wool plug or cap, and sterilise. While the agar is still liquid, slant the tubes sufficiently to leave, when the agar has hardened, a surface about half the length of the tube.

NON–NUTRIENT AGAR:

Add 15 g of non-nutrient agar to 1 litre of modified Neff's amoeba saline (page 16) and bring to a boil, with stirring. Divide into four 500 ml flasks (about 250 ml in each flask), and sterilise. Pour into 9 cm petri dishes to a depth of about 5 mm, flaming the lip of the flask lightly before pouring into each dish. If it is not possible to prepare amoeba saline, use natural or tap water.

Some useful and interesting protozoa: their culture and use

Chilomonas paramecium (Plate 1)

The organism: This flagellate is the most plant-like organism which we shall consider and also one of the easiest to culture. It is a colourless cryptomonad with leucoplasts (colourless rudimentary chloroplasts) rather than the chloroplasts found in its pigmented close relatives.

C. paramecium may be more than 30μm long. It has two flagella of approximately equal length, originating at the anterior end of a deep anterior cavity, the vestibulum. The vestibulum is not a cytostome or mouth, since *Chilomonas* does not take in solid food such as bacteria. Also associated with the vestibulum is the contractile vacuole. The nucleus, with a central nucleolus, is midway back in the cell. The cytoplasm contains many starch bodies, undoubtedly a food reserve. Surrounding the vestibulum are granule-like bodies called trichocysts or ejectisomes, found with the electron microscope to be very unlike the trichocysts of *Paramecium*. Smaller trichocysts are also found in rows under the cell membrane. It has been suggested that these bodies are food reserves, since in starvation they begin to disappear at the same time as the starch bodies, though the reason for their complex structure remains a puzzle.

Chilomonas is common in freshwater habitats where there is a moderate quantity of organic matter. It feeds saprozoically, that is, by absorbing dissolved organic nutrients through its surface. It is of scientific interest especially as one of the 'acetate flagellates', since it needs to be supplied with no nutrients except acetate, ammonium, and thiamine. It can be grown in a chemically defined medium (page 7).

Use in teaching: It is of interest as a flagellate that is easily maintained for microscopic observation and as a food organism for some other protozoa (see *Peranema, Amoeba,* and *Stentor*).

Culture: Chilomonas is commonly grown in polished rice infusion but grows well in any of the grain infusions given on pages 17-18. It can be kept in 50 mm culture dishes, small jars, or tubes and should not require transferring more often than every two or three months. It grows well at sustained temperatures of either 20° or 30°C. If it is used as a food organism for *Amoeba proteus,* a single inoculation into a culture line of the latter is enough, since the *Chilomonas* persists through repeated subculturing. *Chilomonas* can also be grown axenically in a liquid medium (page 19).

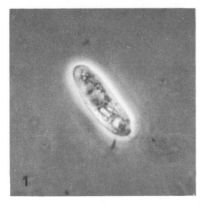
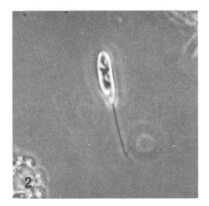

Plate 1   Chilomonas paramecium, × *1,000*
Plate 2   Peranema trichophorum, × *250*
Anterior end toward bottom in both plates. Note that Peranema *is actually much larger than* Chilomonas.

Peranema trichophorum (Plate 2)

The organism: Like *Chilomonas,* this is a colourless phytoflagellate with many pigmented relatives, but it is considerably more animal-like in its nutrition.

Peranema has the slender form common in the euglenids, to which it belongs, and can reach a length of 70 μm. Its body is much more changeable in form than that of *Chilomonas,* though during its steady, gliding advance it does not change shape much. Two flagella come out of the anterior opening of a cavity called the reservoir. However, one flagellum is seldom observed, because it passes posteriorly in a spiral manner, adhering to the pellicle of the organism. The other flagellum is held straight ahead during locomotion, and only the part near the tip beats, reportedly bringing about the gliding motion. This flagellum, which contains a rod as well as the usual microtubules, is thick and easily seen. A contractile vacuole connects with the reservoir near the point where the flagella are attached.

Unlike *Chilomonas, Peranema* feeds holozoically, that is, by the ingestion of solid food. The cytostome or cell mouth has its opening somewhat behind the opening to the reservoir. It is said to take in bacteria and detritus and is known to ingest flagellates of a size near its own, such as *Euglena*. Associated with the cytostome are two pharyngeal rods or trichites, believed to play a role in feeding. *Peranema* can also live entirely on dissolved nutrients, but its nutritional requirements are more complicated than those of *Chilomonas,* that is, there are more substances which it cannot make for itself and must obtain in its food, so that media for axenic culture must be more complex than those for *Chilomonas. Peranema* is thus more heterotrophic, that is, more animal-like in its nutrition, than either pigmented flagellates, which can carry on photosynthesis, or *Chilomonas,* which is colourless but shows much synthetic ability. The food reserve in *Peranema* is the carbohydrate paramylon, characteristic of euglenids, which gives a negative reaction to the iodine test for starch.

The surface of *Peranema* bears a gently spiralling pattern of pellicular ridges, which at best show only faintly with the light microscope. Within the cell, besides the structures already mentioned, the nucleus is usually at the midpoint or a little farther back.

Diagrams of the organism will be found in Mackinnon and Hawes (1961) and Vickerman and Cox (1967). One of the best sources of information about structure, locomotion, and feeding is Leedale (1967).

Peranema often occurs in freshwater habitats with a fairly high content of organic matter and a low level of oxygen.

Use in teaching: These are principally the demonstration of flagellar action and of feeding. Although flagellar action in this organism is very different from that of many flagellates, the ease with which the flagellum is observed makes it particularly suitable. Feeding may be watched after mixing *Peranema* with a good number of *Euglena* or *Chilomonas,* though some time may pass before that event is seen.

Culture: Some suppliers culture *Peranema* in a hay infusion with other flagellates for food, and Mackinnon and Hawes (1961) suggest an infusion of wheat, rice, or peas, with flagellates for food. However, if one can obtain an inoculum of bacteria-free *Peranema*, the axenic method with the easily made *Peranema* medium given on page 19 is highly successful. The medium is poured into 250 ml Erlenmeyer flasks to a depth of about 2·5 cm. Sterile precautions (sterile glassware and flaming of the mouth of the flask) must be observed. Vessels other than flasks can be used if they can be kept sterile. Axenic cultures are just as good at 30°C as at 20°. *Peranema* is usually more abundant in the axenic cultures than in infusion cultures, where one may see many food flagellates before finding a *Peranema.* Even so, this organism never becomes as abundant as *Chilomonas.* We subculture monthly by pipetting. Put a drop or two on a microscope slide to check for presence of the organisms. When removing them from the culture vessel, remember that *Peranema* tends to adhere to surfaces, so it may be necessary to loosen them by swirling or by using the pipette rather vigorously.

Amoeba proteus (Plates 3-6)

The organism: This has been described so often, sometimes at second hand, that no space will be devoted to a description. The reader is referred to such reliable works as Mackinnon and Hawes (1961) and Vickerman and Cox (1967). Abbreviated information and photographs are included in Page (1976). The plates in this booklet are intended to show normal forms and help the teacher identify some of the major characters.

A. proteus, beloved of textbook writers as 'the common amoeba', is not nearly as common as many smaller amoebae, though it is not rare. Its keepers in the laboratory ordinarily feed it on *Chilomonas, Tetrahymena,* or *Paramecium,* but diatoms appear to be a significant part of its diet in nature. It has been reported from habitats with rather low oxygen concentration but far from anaerobic.

Amoeba proteus and one or two other similar organisms are much used in cell biology, and the review volume edited by Jeon (1973) deals in large part with such research.

Use in teaching: It is used to demonstrate the structure of a large cell, amoeboid movement, and phagocytosis, sometimes also contractile vacuole function. Students often have difficulty in observing *Amoeba proteus,* partly because the amoebae are not numerous in any one drop taken from a culture and partly because they do not remain active long under the usual conditions of observation. Care must be taken in dispensing the amoebae to students and in making the slide preparation. It usually works better for the teacher to put a drop on each slide rather than letting several persons dip into the culture. Let the culture stand half an hour so that the amoebae settle to the bottom. Remove them with a pipette, but expel the air from the teat before dipping the pipette into the water, and take care not to stir up the culture. After putting a drop of the culture on a slide, add a cover slip supported by a thin ridge of petroleum jelly along each of two opposite sides. This prevents crushing and at the same time leaves two sides open for gas exchange. Trapping an air bubble under the cover slip will provide a useful source of oxygen to keep the amoebae active longer. If the preparation is not allowed to dry out (water can be added at the open sides of the preparation), it should be usable for a half hour or more. The amoebae take a few minutes to begin locomotion. When first subjected to light from the microscope lamp, an amoeba will stop but will soon resume movement. By manipulation of the slide or partial shading of the substage illumination, one may observe what appears to be an attempt to avoid the bright light. This phenomenon has been observed by many persons.

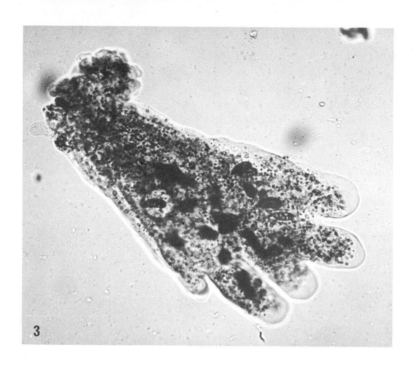

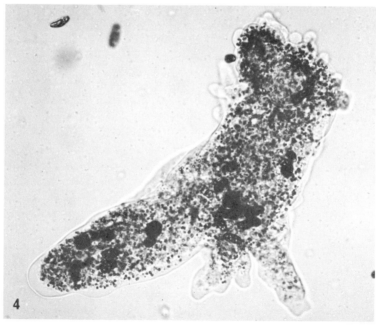

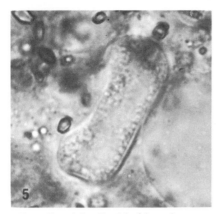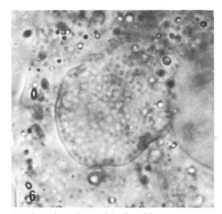

Plates 5 and 6   Nuclei of Amoeba proteus, *showing side view with discoid appearance (Plate 5) and full view (Plate 6). Notice the crystals, which are triuret, a nitrogenous waste product, × 1,000.*

A Polish worker (A. Grebecki) has recently concluded that increased light stimulates contractile activity in the illuminated region, forcing the endoplasm to flow toward the shaded regions where contractile activity is lower.

Culture: This method, not original to us, has proved satisfactory for easy maintenance of cultures for teaching, assuming that the inoculum includes not only *Amoeba proteus* but also a food organism such as *Chilomonas paramecium.* Pour Prescott's and James's solution (page 16) into a low dish such as a crystallising dish or 'finger bowl' (diameter about 100mm) to a depth of 1 cm or slightly more. Add 3 or 4 uncooked grains of polished rice. If necessary, use natural water of the quality advised (page 16) for grain infusions rather than Prescott's and James's solution. If possible, let the rice grains stand in the water for two or three days before adding amoebae. Inoculate about 2 ml of a healthy culture of *A. proteus* containing the food organisms. Cover with half a petri dish or other loose cover. Both evaporation and contamination may be reduced if the vessel is closed, either inside or outside the cover, with clingfilm. Keep at 18° or 19°C if possible. Although some workers report growth at temperatures up to 30°C, we have not been able to get growth or even survival with prolonged exposure to that temperature, though the cultures may survive day-time temperatures somewhat above 20° if the nights are cooler.

◀ *Plates 3 and 4   Amoeba* proteus, × 250. In Plate 3, the polypodial amoeba is moving toward the lower right, with the uroid at upper left. In Plate 4, the amoeba, essentially monopodial with the pseudopodia at lower right being withdrawn, is moving toward the lower left.

In six or eight weeks the culture should contain good numbers of amoebae, including some amongst the fungal mycelia usually growing from the rice grains, and the cultures should still be usable three months after inoculation. In a healthy culture, amoebae are moving with pseudopodia well extended. When most amoebae are monopodial elongated forms (only one pseudopodium), the culture is no longer in very good condition. Rounded cells are often unhealthy, particularly smoothly rounded ones with no pseudopodial projections.

Since not all cultures inoculated will be successful, several should be started at a time. Subculture every two months. *A. proteus* seems to need a few bacteria, and sterilisation of the culture dishes is not recommended. The chapter by J. L. Griffin in Jeon (1973) is an essential reference for those who wish more refined methods or larger numbers of organisms than may be provided by the method suggested here.

Saccamoeba limax (plate 7)

The organism: This freshwater amoeba is not rare in the Northern Hemisphere. It has an average length of about 55µm, sometimes up to 75µm, and is thus considerably smaller than *Amoeba proteus* but still suitable for observation by students. Its advantages for teaching include abundance in culture and easy observation of nucleus, contractile vacuole, and cytoplasmic streaming. However, for best results, it requires somewhat different handling from *A. proteus.*

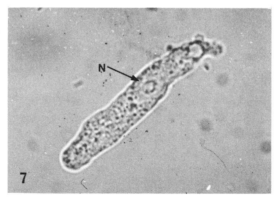

Plate 7 Saccamoeba limax, × *1,000. The amoeba is moving toward the lower left, with the uroid and, just anterior to that, the contractile vacuole at upper right. N = nucleus.*

Saccamoeba moves as an unbranched cylindrical body, though it may put out a temporary branch to change direction. Inactive amoebae are irregularly shaped. The nucleus is especially prominent, with a central nucleolus separated by an optically clear space from the nuclear membrane, which can be discerned at higher magnifications. At the posterior end is the uroid, which may be bulbous and often bears many temporary, fine, hair-like projections or villi (not cilia). The contractile vacuole empties with noticeable bulging at, or just in front of, the posterior end. The cytoplasmic crystals are of the form shown in larger amoebae to be triuret, a nitrogenous waste product.

Saccamoeba limax and some other species do not form cysts, but at least one species of this genus does encyst.

Culture: This must be explained before the method of harvesting for use in teaching can be understood. Although *S. limax* is of aquatic origin, it grows best on non-nutrient agar (page 19) without a liquid overlay. The non-nutrient agar in a petri dish is streaked with *Escherichia coli* removed from a slope of nutrient agar (page 19) with a bacteriological loop. The amoebae are then added either as a cell suspension in modified Neff's amoeba saline (page 16) or or on a small (2 or 3 mm square) block of agar cut from a parent culture and turned onto the fresh agar surface with the original culture surface, bearing the amoebae, down. Thousands of amoebae can be obtained on a few agar plates in a week or ten days. However, if it is inconvenient to make up agar plates, *S. limax* can be cultured in polished rice, barley, or maize infusion in 50 mm culture dishes. A loopful of *Escherichia coli* can be added to the infusion, or the bacteria from the seed culture can be allowed to grow in the infusion for a few days before the amoebae are added. This amoeba grows well at 20°C but did not survive prolonged exposure to a temperature of 30°C. However, our cultures are often exposed to day-time temperatures a few degrees above 20°. Subculture every three or four weeks if on agar. If in liquid, some experimentation may be needed to determine the interval between transfers.

Use in teaching: These amoebae are especially suited for demonstration of cytoplasmic streaming during locomotion because of the simplicity of their form and the convenient markers provided by the crystals. Observation of the contractile vacuole is easy because of its position unobscured by cytoplasmic inclusions. The emptying process can be studied and the period from one systole to the next timed without difficulty.

Amoebae grown on agar must be washed off with a little liquid to prepare for observation and can thus be concentrated. Use a sterile pipette, lifting the petri dish cover as little as possible, and wash the agar surface gently with no more than 1 ml of sterile modified Neff's amoeba saline, natural water, tap water, or distilled water. The process can be repeated two or three times. One ml of suspension should yield nearly 20 drops for slides or hanging drops.

Amoebae may be observed on a slide, preferably with the coverslip supported on two sides by petroleum jelly as suggested for *Amoeba proteus*. However, the amoebae soon round up and cease activity under those conditions, especially if taken from an agar culture with a fairly high number of bacteria. An air bubble trapped under the coverslip will extend the usefulness of such preparations, and the amoebae will be seen to orient themselves toward the bubble. It is recommended that, if there is time, hanging drops should be prepared, using either thick slides with a concave depression ground into them or slides with an attached ring to support the coverslip 2 or 3 mm above the slide. Put petroleum jelly on the slide or attached ring to join it to the coverslip. Put a single drop of cell suspension in the centre of a coverslip and place the slide, concavity or ring down, on the coverslip, pressing gently to assure even contact. After about an hour, many amoebae will have attached to the coverslip and the slide can be turned over, avoiding any running of the drop, to leave the drop hanging and the amoebae moving on the underside of the coverslip, where they can be observed under any magnification. The preparation should be good for hours. After a half hour, many of the bacteria will have fallen from the underside of the coverslip, and the amoebae should then be moving about more freely. This method can be used for many other medium-sized and small amoebae, but not all attach well. If desired, a coverslip bearing a hanging drop can be transferred to a flat slide for observation by such optical systems as phase contrast. However, it should not be pressed down hard on the flat slide, and on a flat slide the preparation cannot be expected to remain usable as long as if it had been left hanging.

If *Saccamoeba* is cultured in liquid, the drops from which the hanging drops are prepared may have lower populations of amoebae than those made by washing an agar surface with small quantities of liquid. However, the numbers should still be far higher than those in preparations of *Amoeba proteus*.

Naegleri gruberi (Plates 8-11)

The organism: This is the amoeboflagellate most often reported in the Northern Hemisphere and is also known from the Southern Hemisphere. Like other amoeboflagellates, it has three forms: amoeba, flagellate, and cyst. It is valuable for teaching because it can be cultured easily, even at temperatures above those usual in northern Europe, and especially because all three forms can be obtained easily and their transformations observed.

The amoeboid stage (about 15-30 μm long) is the only one in which *Naegleria* feeds and divides. Locomotion is much more eruptive than in *Saccamoeba*, with hemispherical bulges of clear cytoplasm formed to either side of the anterior end. Plate 8 shows the active locomotive form, but the amoebae are often more irregular in shape. The nucleus, with a central nucleolus, is often surrounded by a conspicuous ring of granules. The contractile vacuole usually empties at the posterior end, which may trail several uroidal filaments produced by adhesion to the substratum.

The flagellate phase, easily induced by the method suggested below, is temporary. Although *Naegleria* is normally biflagellate, cells with more flagella are often found. The nucleus is at the anterior end of the flagellate, the contractile vacuole near the posterior end. Unlike some amoeboflagellates, *Naegleria* does not have a cytostome.

The cysts are more or less spherical, with a smooth wall made up of a thin outer layer and a thicker inner layer, and containing several plugged pores.

Two good sources of information on *Naegleria* and other amoeboflagellates are the reviews by Fulton (1970) and Schuster (1979). Information on the lethal pathogen *Naegleria fowleri* will be found in the review by Griffin (1978); the subject is also discussed by Schuster (1979). There is no reason to fear or hesitate to use the common species *Naegleria gruberi* in teaching. Those who intend to isolate their own strain of *N. gruberi* are referred to the section Isolation from natural sources (page 45-46) for suggestions on confirming the identity of their isolates.

Culture: Naegleria gruberi is usually cultured on non-nutrient agar with *Escherichia coli* for food, as described for *Saccamoeba limax.* It grows and encysts well at either 20° or 30°C. The cysts remain viable for months on agar plates if the agar does not dry out to a hard film. On agar slopes at 4° to 9°C they remain viable for several years. *N. gruberi* can also be cultured in barley, polished rice, or maize infusion, feeding on added *E. coli* or on bacteria accompanying the inoculum of amoebae.

Use in teaching: The amoebae are best observed in hanging drops as suggested for *Saccamoeba.* Transformation to the flagellate stage is induced by adding about 1 ml of liquid (modified Neff's amoeba saline or natural, tap, or distilled water) to an abundant young culture on agar, preferably one inoculated 24 to 36 hours earlier from an abundant parent culture. Wash the surface thoroughly but gently to obtain a large number of cells, and prepare hanging drops, which can be turned over within an hour. Flagellates should begin to

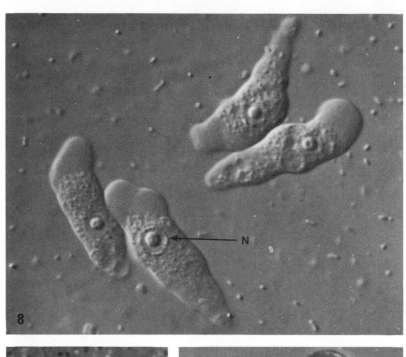

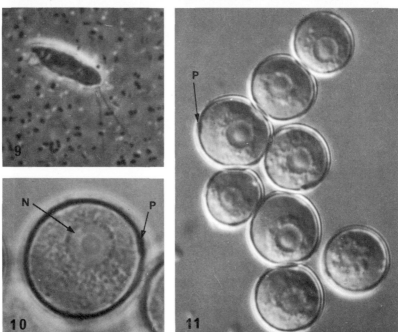

appear within two hours, often sooner, with a peak at about four hours (temperature about 23°C), after which flagellates will begin to transform back to amoebae. It is possible to watch amoebae moving on the underside of the coverslips in the same preparations. In practice, one usually obtains flagellates in the course of setting up hanging drops for observation of amoebae.

To examine flagellates more closely and to count flagella, put a drop of saturated mercuric chloride solution on a flat slide and add the coverslip bearing the hanging drop. Nuclei and sometimes basal bodies of flagella are made conspicuous with this fixation of wet mounts. In laboratories where Lugol's iodine solution is kept on hand for other purposes, it may be substituted for mercuric chloride solution.

Encystment is readily observed. Within 48 hours of inoculation on to a fresh agar surface, early cyst stages begin to appear, and in another day or two many cysts are present. Excystment is more easily observed in *N. gruberi* than in most amoebae. Prepare a suspension of cysts by washing the agar surface with liquid. Put a drop or two of suspension on to a slide, and place the slide, without a coverslip, into a moist chamber made by moistening a disc of filter paper or other absorbent paper in the bottom of a petri dish or other shallow container. After two or three hours the slide can be removed from the moist chamber, a coverslip added, and observations made. With this species one may easily see several cells excyst within a short time. Amoebae of some species excyst by digesting or breaking a hole in the cyst wall; the rapid excystment of *Naegleria* may be due to its method of exit through a preformed pore.

These methods can be modified for liquid cultures. Concentrating the cells by pipetting or centrifugation, then substituting fresh liquid, should provide an adequate stimulus for either flagellate transformation or excystment. One sometimes sees flagellates formed spontaneously in liquid cultures.

Actinophrys sol (Plates 12, 13)

The organism: This is the most commonly found heliozoan in the Northern Hemisphere, is easily cultured, and provides a good introduction to the

---

◄ *Plates 8-11* Naegleria gruberi. *In Plate 8 (differential interference contrast, × 1,000), notice the hyaline pseudopodial bulges and the nuclei (N) surrounded by a ring of conspicuous granules. In Plate 9, the flagella of this flagellate stage (× 1,000, fixed with HgCl₂) are at lower right. Plates 10 (bright field, × 2,000) and 11 (differential interference contrast, × 1,500) show cysts. P = plugged pore.*

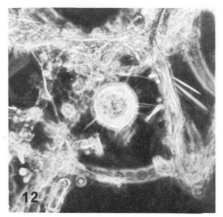
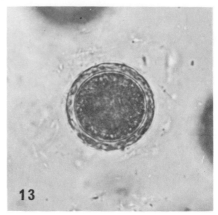

Plates 12 and 13   Actinophrys sol. *Plate 12 (anoptral contrast, × 250) shows an* Actinophrys *floating amongst aquatic vegetation; Plate 13 (× 2,000), an oöcyst formed in culture.*

heliozoans (little sun animals) with their fine pseudopodia radiating from a spherical body. Although a few heliozoans are attached to the substratum by a stalk and some unattached ones can move over surfaces, they typically float in the water and use their pseudopodia to feed.

Actinophrys sol is spherical, with a diameter usually near 40 to 50 μm, though there may be considerable variation. The pseudopodia, which radiate in all directions, are called axopodia because they contain a central core or axoneme, which can be discerned with the light microscope, given proper magnification and illumination. The electron microscope shows that the axoneme is a complex structure made up of microtubules. The cytoplasm can be divided, though not very sharply, into an outer, somewhat foamy appearing ectoplasm and a less vacuolated endoplasm. At the centre is a single nucleus; its nucleolar and chromosomal material are arranged as a layer just inside the nuclear membrane, so that its central area appears more or less empty. The axonemes are attached to the outer surface of the nuclear membrane. On the outer edge of the cell, the contractile vacuole bulges far above the surface. This is a naked heliozoan, that is, one without scales or spines.

In cultures where the *Actinophrys* have become abundant, one may eventually find numbers of cysts (Plate 13). These are not, however, ordinary 'resting' cysts like those formed by *Naegleria gruberi*. Rather, they are, when mature, the products of a sexual process called paedogamy, resembling autogamy in ciliates. A single cell encysts, after disappearance of the axopodia. A progamic division produces two gamonts, which then undergo two maturation divisions, followed by cytoplasmic and nuclear fusion. The cyst, which at first had a thin covering, toward maturation becomes covered with overlapping scales. This is an oöcyst.

Use in teaching: Although its first use will be as an example of heliozoan and axopodial structure, *Actinophrys* provides especially interesting material for the observation of feeding, either in a hanging drop or a shallow dish containing both predator and prey. If the cultures can be brought to a high population density by generous feeding, it may be possible to follow the sexual process described above.

Culture: Our best results have come from an adaptation of the method used by Professor J. A. Kitching for *Actinosphaerium.* We use his *Actinosphaerium* medium (page 16), but natural water prepared as described on page 16 is also satisfactory. Although we aerate the medium (page 16) for all heliozoans, that step may be omitted for *Actinophrys.* Pour fresh medium into a 50 mm culture dish or equivalent vessel to a depth of about 1 cm. Add a pipette full of *Actinophrys* suspension from a thriving culture, then a pipette full of *Colpidium striatum* from a dense culture (page 36). If possible, centrifuge the *Colpidium* and replace the original culture liquid with fresh natural water or *Actinosphaerium* medium before adding the ciliates to the *Actinophrys* culture. An established culture should be fed once or twice a week with about a half pipette full of *Colpidium.* The cultures can be kept a long time. We start new cultures every week, but with adequate feeding and occasional removal of excess liquid that interval can be prolonged to several weeks.

This same culture method is suitable for *Actinosphaerium,* which is larger than *Actinophrys* but does not become so abundant and is more likely to die off, and for the scale-bearing heliozoan *Raphidiophrys.* It will also serve for *Acanthocystis,* which we ordinarily feed with *Tetrahymena* washed by centrifugation.

Tetrahymena (Plate 14)

The organism: According to Corliss (1979), more than 3000 papers were written 'about' *Tetrahymena* in less than 40 years after the generic name was coined by Furgason in 1940. That figure does not appear to include publications in which these organisms figured under another generic name between its first axenic cultivation by Lwoff in 1922 and erection of the genus in 1940. Because it is easily cultured in large numbers and has other interesting characteristics, *Tetrahymena* has become one of the most popular subjects for

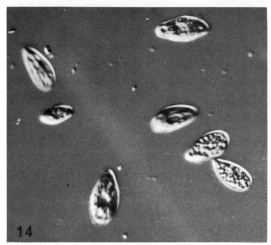

Plate 14 Tetrahymena thermophila *in axenic culture.*
Differential interference contrast, × *250.*

biochemical and cell biological work. Readers who wish more information than is given here and a gateway to the large literature should consult Elliott (1973), Nanney and McCoy (1976), and relevant pages in Corliss (1979).

The *'Tetrahymena pyriformis'* of the literature has now been divided into 14 species (Nanney and McCoy, 1976), a proposal which has gained wide acceptance.

The organisms formerly included in *T. pyriformis* in the broad sense are pear-shaped hymenostome ciliates, with a mean length often about 50 μm (but with much variation). The body cilia are organised into rows and are of much the same sort all over the body except in the mouth. The mouth or cytostome is located in a buccal cavity a little way behind the anterior end, which is the more pointed end of the ciliate. Within the mouth are four compound ciliary structures, an undulating membrane on one side and three membranelles on the other; special techniques are needed to examine these. The contractile vacuole is in the posterior part of the cell. Ciliates have two kinds of nuclei, a polyploid macronucleus, which regulates vegetative functions, and a diploid micronucleus, which is involved in sexual processes such as conjugation. *Tetrahymena* has a central, oval macronucleus. The micronucleus, when present, lies in an indentation of the macronucleus. However, many of the laboratory strains lack a micronucleus and are therefore unable to conjugate. All the strains which Nanney and McCoy retained in their more restricted *T. pyriformis* are amicronucleate, as are three of their new species.

Our illustration (Plate 14) gives only an idea of the general form of *Tetrahymena*, similar perhaps to the fleeting glance with which many workers concerned with non-morphological aspects are content.

Tetrahymena is widely distributed in freshwater habitats, where it usually feeds on bacteria.

Use in teaching: The wide use of *Tetrahymena* in all sorts of experimental work in itself justifies its inclusion here. It is especially suited for any investigations requiring a single species with no contaminant organisms. These include not only biochemical studies but experimental ecology such as investigations of population dynamics and predator/prey relationships, since *Tetrahymena* feeds on bacteria and can itself serve as prey for larger protozoa.

Culture: Tetrahymena's nutrition has been much investigated, and it can be grown in chemically defined media. However, we use a more complex medium, proteose peptone/yeast (page 19). We usually subculture every five weeks, leave them a few days at room temperature (ca. 22°C), and then store them at 13-15°. Subculturing is done simply by flaming the mouth of the tube containing the culture and that of the tube containing fresh medium lightly, then pouring a millilitre or two from the first into the second, flaming again, and closing the tubes. The five-week interval between transfers is on the cautious side; the cultures can be kept for at least two or three months at 4°C between transfers. In our experiments, cultures of a laboratory strain now classified as *T. thermophila* appeared only slightly if at all less abundant after nine days at 30°C than those kept at 20°. But a strain still classified as *T. pyriformis* in the sense of Nanney and McCoy did much better at 20° than at 30° over the same period. The comments made earlier (page 10) about the shortening of culture life at higher temperatures should be kept in mind.

If axenically cultured *Tetrahymena* are to be fed to other protozoa such as *Amoeba proteus*, they should be washed free of the proteose peptone by centrifugation and resuspension in water or a weak organic salt solution (such as Prescott's and James's) at least three times. Otherwise the proteose peptone may produce an unwanted increase in the bacterial population accompanying the predator.

Colpidium striatum (Plate 15)

The organism: This ciliate belongs to the same family as *Tetrahymena* and has similar oral structures. It is about 85 μm long, more elongate than *Tetrahymena*, and somewhat reniform, with the position of the mouth indicated by a depression about a quarter of the way back from the anterior end.

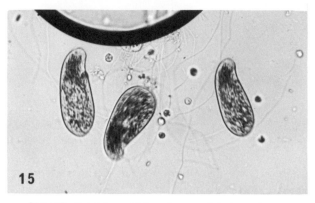

Plate 15 Colpidium striatum *from an infusion culture; the crescent at the top is the edge of an air bubble.* × 250.

Use in teaching: Colpidium is included here mainly because it thrives so well in culture with bacteria. It is very suitable for experimental work on feeding relations and is an excellent food organism for certain other protozoa, such as heliozoans and suctorian ciliates. Several workers have remarked that it appears to give better results as a food organism than does *Tetrahymena* with some of the same predators.

Culture: For feeding to heliozoans and suctorians, we maintain *Colpidium* in 150 ml flasks containing *Actinosphaerium* medium (to a depth of 2 or 3 cm) and about four boiled barley grains. Simple infusion cultures in 50 mm culture dishes are so successful that they are recommended as the routine method. *Colpidium striatum* grows well in all the grain infusions described on pages 17-18 least well in lawn grass seed infusion. Cultures with barley or maize are especially dense. Cultures with millet grow well but decline sooner than those with some other grains. Infusion cultures with barley, maize, wheat, and rice remain abundant for two months or longer. At the CCAP our 'permanent' cultures for maintaining the strain and supplying orders are kept in beakers or tubes of biphasic soil extract with barley grains (page 18). They are subcultured every four weeks, and to each tube are added one or two drops of a milky-looking suspension of baker's yeast. However, soil extract cultures seem unnecessary for teaching purposes because the infusion cultures are so long lasting.

The one difficulty with *C. striatum* if a European isolate is used may be temperature, since cultures kept at 30°C for little more than a week contained no organisms, while parallel cultures at 20° thrived. However, a reduced night-time temperature may permit survival even if day-time temperatures are somewhat elevated.

Paramecium caudatum (Plates 16-18)

The organism: 'Paramecium species are certainly the most widely spread ciliates in the world', according to E. Vivier in the most recent book devoted entirely to this genus (Van Wagtendonk, 1974). It is, at least in the Northern Hemisphere, easily cultured from various bodies of fresh water.

Many textbooks contain labelled illustrations of *Paramecium caudatum*. Its general resemblance to a shoe sole or slipper was noticed almost as soon as it was seen. Its length may be from 180 to 280 μm. The anterior end is rounded, the posterior end conical. A conspicuous, slightly curving oral groove leads to the buccal cavity approximately in the middle of the body. The cytostome or mouth, at the posterior end of the buccal cavity, is the place where food vacuoles are formed. (Like *Tetrahymena* and *Colpidium*, *Paramecium* is a hymenostome, though its oral ciliature differs somewhat from theirs.) *Paramecium*, like other ciliates, has a cytoproct or pore through which wastes remaining after digestion are eliminated; it is about half way between the cytostome and the posterior end. There are two contractile vacuoles, one anterior and one posterior, with easily observed ampullae emptying into them, and each contractile vacuole empties to the outside through a definite and permanent pore. Approximately in the middle of the cell is the large ellipsoid macronucleus, discernible in the living animal if one has an idea of its appearance. The micronucleus, in an indentation in the macronucleus, can be found only after staining. Finally, there is a layer of rod-like trichocysts all around the periphery of the cell, just beneath and perpendicular to the pellicle. The function of the trichocysts of *Paramecium* remains obscure but is not defensive.

Use in teaching: Paramecium caudatum is the generally favourite ciliate for teaching purposes. Except for observations of locomotion and of the avoidance reaction, one soon encounters the problem of keeping it in one position long enough. The best solution is to find individuals feeding in one place where food is abundant enough to keep it from moving away, for example, amidst detritus or near a grain. Several agents have been used to slow *Paramecium* and other ciliates. Methyl cellulose (10 per cent in water) or sodium carboxymethylcellulose (2 per cent) increase the viscosity of the preparation. With a thin stick, put a ring of methyl cellulose on a slide, leaving in the middle of the ring a space large enough for a drop of *Paramecium*

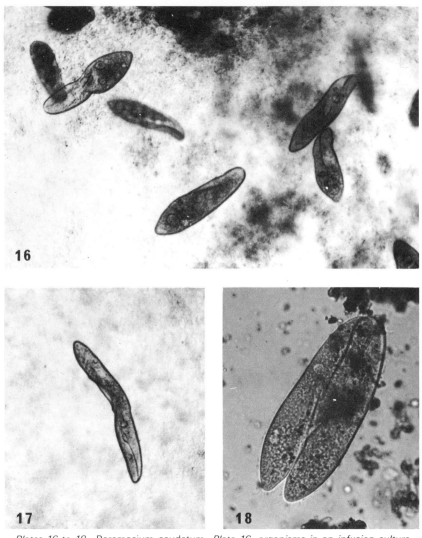

Plates 16 to 18 Paramecium caudatum. *Plate 16, organisms in an infusion culture, including one dividing* Paramecium *at upper left. The oral groove is seen distinctly in several organisms.* × 100. *Plate 17, another fission in an infusion culture.* × 100. *Plate 18, conjugation in mixed material from a pond.* × 250.

suspension. Then add a coverslip, pressing it down gently but not squashing the ciliates. The concentration of methyl cellulose will be thicker in some parts of the preparation than in others. It is said that methyl cellulose can affect contractile vacuole function, one of the phenomena which one usually wishes to observe. A very weak solution of nickel sulphate (kept as a 0·1 per cent stock solution of $NiSO_4 \cdot 6H_2O$ and diluted to 0·001 per cent) has sometimes been used to slow movement, but this is said also to affect contractile vacuole function.

Digestion is one process which can be followed in *Paramecium*. Baker's yeast is stained with the vital dye Congo red (about 0·1 per cent solution in water) and added to the *Paramecium* on a slide. Congo red is an indicator which turns blue as conditions become more acid, and colour changes can be observed as the vacuoles circulate in the cytoplasm. However, the correlation of these colour changes with the actual digestion is uncertain. Morphological changes in the ingested food organisms may provide some guidance.

The use of a solution of the dye methyl green in weak acetic acid (1 g of methyl green in 100 ml of 1 per cent acetic acid) makes two kinds of observations possible. After a drop of *Paramecium* culture has been covered with a coverslip, add a drop of methyl green/acetic at one side of the coverslip and tilt the slide so that the stain mixes with the organisms. The methyl green will stain the nuclei, and the acetic acid will cause extrusion of the trichocysts. However, where the mixture is too dilute, the nuclei will not be stained, and where it is too concentrated, the stain may be too heavy and the cells distorted. Students may also be able to observe the avoidance reaction of *Paramecium* at the interface between water and weak acid as the latter gradually advances.

Transverse fission will be seen in rapidly multiplying cultures. Conjugation, during which nuclear material is exchanged, can be observed when individuals of different mating types are present, as in intentional mixtures of two strains or in mixed material from a natural source. Conjugation is favoured by high population density or by the onset of a period without food after a time of plentiful food.

Culture: Some methods for culture of *Paramecium* are discussed by Vivier in Van Wagtendonk (1974). At the Culture Centre of Algae and Protozoa, we grow *P. caudatum* in 250 ml beakers containing biphasic soil extract with barley grains (page 18), starting new cultures every five weeks. These cultures generally remain usable for months; they are cleaned every two weeks by removing excess detritus, and at that time three boiled barley grains are added to each beaker. Baker's yeast, rolled into a ball the size of a pea, may be useful in getting new cultures started. The ciliates may be found near the top of the liquid or around the grains. They can also be grown in tubes of the same medium, adding one or two drops of a milky-looking suspension of baker's yeast at the time of inoculation.

Infusion cultures can be maintained in 50 mm culture dishes. Wheat gives good results, as does polished rice, though cultures with the latter grain appear to decline sooner. The cultures with maize are not as good as those of some other ciliates, in our experience, and millet cannot be recommended. I have for years maintained *P. caudatum* in lawn grass seed infusions. *Paramecium* takes longer than *Colpidium* to become numerous in infusions. New cultures should be started monthly. Our strain did not survive prolonged exposure to 30°C, but other workers report successful cultures at temperatures from 0° to 30°. Readers might find that a few attempts to isolate the organism from nature would give them a suitable indigenous strain, although it might be possible to adapt European strains gradually to higher temperatures.

Stentor coeruleus (Plate 19)

The organism: No one who has seen the blue *Stentor* will doubt that it is among the splendours of the microscopical world. This heterotrich ciliate is spectacular for its trumpet-shaped attached form, between 1 and 2 mm long; for the blue colour localised in stripes just beneath the pellicle; and for the crown of membranelles, compound ciliary structures, which spirals clockwise around its broad end and down to the mouth, giving the illusion of wheel-like

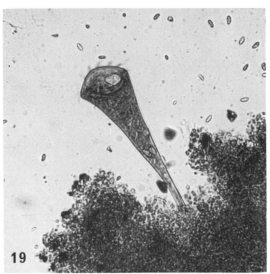

Plate 19 Stentor coeruleus *from a culture, extended and attached form. Note adoral membranelles (blurred because moving) around oral end.* × 100.

40

movement while creating currents that carry food organisms to the mouth. *Stentor* is large enough to be seen as a blue speck with the naked eye and to be examined in detail under relatively low magnification.

The form when attached is well shown in Plate 19. Besides the adoral zone of membranelles, the mouth, and the pigment stripes, it should be easy to find the contractile vacuole, a little below the mouth, and the long, beaded macronucleus. The whole body is covered with cilia. *Stentor* is highly contractile, with myonemes or bands of contractile fibres running longitudinally beneath the pellicle. It is not permanently attached, and one often sees it swimming about in a more contracted, oval or barrel-shaped form.

Use in teaching: Stentor is of interest first of all simply as an organism whose appearance and activities are fascinating to watch, illustrating so well the potentialities of the unicellular organism. Since it feeds on bacteria, algae, smaller protozoa, and rotifers, it is a good subject for experiments to study food preference. For anyone who wants to try something difficult, it is large enough to be of special value for microsurgical procedures, particularly as a method of investigating morphogenesis. There is a brief discussion of its use for that purpose in Sleigh (1973) and more detailed discussions by the leading experimental worker with *Stentor* in Tartar (1961) and Tartar's chapter in Chen (1967b).

Culture: Although wheat and polished rice infusions were satisfactory and lawn grass seed proved usable for short term cultures, maize kernels are especially good material and should be used if they are available. In maize infusion (one kernel in a 50 ml dish), some *Stentor* can always be found swimming about, but large numbers are attached at some points to the kernel, which to the naked eye sometimes appears covered with a bluish fuzz. These attached organisms are easily dislodged. Maize gives not only high population density but also relatively long culture life, and subculturing every two months should be adequate for maintenance. One can not only transfer organisms to a dish containing a freshly boiled kernel but also transfer the kernel from the old culture to a dish of fresh water to get increased numbers.

In our laboratory *Stentor* is grown in large numbers in beakers of biphasic soil extract with barley grains (page 18). New cultures are started monthly, but the cultures remain good for several months if cleaned and fed every two weeks. This process consists of forcing the attached *Stentor* off the sides of the beaker by directing water at them forcefully with a pipette; wiping the inside wall of the beaker to make a cleaner surface for attachment; removing detritus and old grains from the beaker; adding fresh boiled barley grains; and topping up the water if necessary. *Spirostomum ambiguum*, another highly contractile heterotrich, but so long (up to 3 mm) and slender that it might be taken for a worm, is grown in beakers in the same way, except that the wall of the beaker does not need to be wiped, since it does not attach.

Our European strain of *Stentor* does not survive prolonged exposure to a temperature of 30°C. For those who wish to isolate an indigenous strain, *Stentor* is often found in habitats of rather low oxygen concentration with a moderate quantity of organic matter. *Spirostomum* appears to have an even greater affinity than *Stentor* for low oxygen concentration, and the bottom of the culture beaker is often alive with masses of these worm-like ciliates.

Vorticella microstoma (Plate 20)

The organism: Vorticella is a genus, with more than 200 species, of sessile bacteria-feeding ciliates. It is representative of a group of hymenostomes

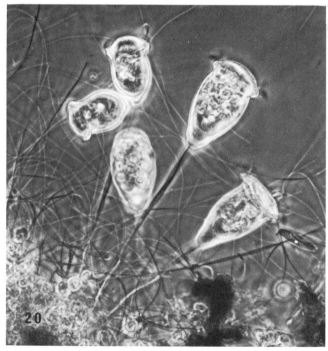

Plate 20   Vorticella *sp. Several individuals in material collected from a river. Note the slightly spiralled spasmoneme in the stalk of the second individual from the right.* × *250.*

called the peritrichs. *Vorticella* is common in freshwater habitats with adequate bacterial food, including sewage sludge.

The mature form consists of a zooid or individual, shaped like an inverted bell, and a stalk which at its lower end is attached to the substratum. The zooid, about 35 to 80 μm long in various species, lacks body cilia, though faint pellicular striations run around it. Encircling the broad upper end of the zooid are two compound rows of cilia, said to be homologous with the oral structures of other hymenostomes. These are involved in feeding; it is believed that some of them set up the currents that bring in the bacteria and others pass the bacteria on to the mouth. This buccal ciliature winds in an anti-clockwise direction to the mouth, which is at the bottom of an infundibulum, into which the contractile vacuole also leads. The long, curved macronucleus is easily seen. Within the zooid are myonemes running both longitudinally and around the circumference; when they contract, the bell-shaped zooid becomes more or less spherical, with the upper end and its feeding organelles drawn in. In the stalk is a spasmoneme, seen in Plate 20 as a slightly spiral thread. When it contracts, the entire stalk takes on a helical shortened form.

Reproduction involves the production of the swimming stage or telotroch. When the zooid divides, one daughter cell remains on the stalk. The other produces toward its posterior end a ring of cilia, swims around, and settles down. The trochal cilia are lost, and the scopula, a special ciliary structure at the posterior end, produces the outer part of the stalk. A few days after a culture of this species has been inoculated and has developed into a fairly dense population, the third form, the cyst, begins to appear. *V. microstoma* encysts readily.

While *Vorticella* is solitary, that is, with one zooid per stalk, many individuals on their separate stalks may be clustered together. Many peritrichs are colonial, with branched stalks.

Use in teaching: Vorticella is a favourite object for the study of contractility. One need only tap the table or the microscope stage to see both contraction of the zooid to form a spherical shape and contraction of the stalk into a helical form. The contractile proteins in peritrich stalks have been investigated biochemically. Feeding can be observed with demonstration of the feeding currents by movement of particles. In *V. microstoma,* all three stages of the life history are easily obtained. If one inoculates a new culture from an old one containing zooids and cysts or cysts alone, one can usually find a fair number of teletrochs the next day.

Detailed suggestions for the use of *Vorticella* in teaching have been published by A. R. Jones (1980), who with good justification considers it more suitable than *Paramecium*.

Culture: Vorticella microstoma is unique amongst the ciliates discussed in this publication (other than *Tetrahymena thermophila*) in that it grows well at 30°C, often better than at 20°. It is therefore well suited for use in warm countries. In our tests, barley, wheat, and polished rice infusions gave good results at 30°C. At 20° good results were obtained with polished rice and acceptable results with wheat. Maize proved very successful at 20° and 30°. As with *Stentor,* the ciliates settled on the maize kernel, until it was covered with many *Vorticella.* Eventually, many cysts were present, both on the kernel and on the bottom of the culture dish. With maize infusion cultures, transferring every two months is sufficient to keep the strain going, but after a couple of weeks many of the organisms have encysted. To obtain large numbers for study, it may be desirable to make successive subcultures a few days apart if one is starting from cysts. To keep active zooids, we subculture every week, and with species which do not encyst as readily as *V. microstoma* it is best to subculture every week or two until one is certain that they will last long enough to lengthen the interval. Transferring a maize kernel from a culture as old as five months to fresh liquid yields zooids of this encysting species of *Vorticella* in a day or two.

Those seeking their own isolates from nature should have no difficulty finding *Vorticella,* which often appears in collected material treated according to the procedure described in the following section.

Isolation from natural sources

Once a little experience in culturing protozoa has been gained, isolation from nature will prove much less difficult than identifying the organisms isolated.

In general, ciliates grow out (become numerous enough to be noticed) well in infusions. One may inoculate some collected material into an infusion or add a few boiled grains to the collected water. The resulting mixed culture can be observed over a period of weeks, since a succession of species will appear and disappear. Generally, if one wishes to isolate a particular organism, one should do so as soon as it appears, since other organisms may soon replace it. Obviously, all these species were present in collected material, but the time when they become noticeable and then disappear depends on the rapidity with which they multiply, their interrelationships as predator and prey, and the changing physical and chemical conditions in the mixed culture. If a protozöon is present, then logically its food must also be present. If it is feeding on bacteria, one can expect to transfer some of the food bacteria along with the protozöon. Nevertheless, one will usually have more failures than successes and should not be discouraged from further attempts.

As mentioned previously small and medium-sized amoebae are often cultured on agar. Collected material may be inoculated on to an agar made with one of the grain infusions (but very few grains actually in each dish), and the amoebae can later be transferred to non-nutrient agar streaked with bacteria or kept on fresh grain-infusion agar. Methods are given in Page (1976).

Readers especially in warm countries should be aware of the possibility of isolating the lethal pathogen *Naegleria fowleri*, which grows at temperatures up to 45°C but may be present in waters at lower temperatures within the range of day-time temperatures in some countries. However, these lower temperatures (even in the low 30's) favour non-pathogenic amoebae including *Naegleria gruberi*, so that there is little likelihood of isolating the pathogenic *N. fowleri* if the cultures are kept at 30° or below, and even in the range up to 37° or more the non-pathogens are likely to overgrow *N. fowleri*. Morphologically, *N. gruberi* can be distinguished from other species of the genus, especially from *N. fowleri*, by the fact that the pores in the cyst wall have noticeably thickened, collar-like rims.

Protozoa feeding on algae may grow out in the soil extract (E + S) medium left in the light (not direct sunlight).

Soil as well as freshwater organisms can be cultured in infusions or on agar.

No marine organisms are included among the species for which detailed culture instructions are given. In general, marine ciliates will grow well in Føyn's Erdschreiber (page 18) if their food is present. Some marine protozoa feed on algae, which may grow in Erdschreiber, but even in that case the cultures should not be kept in direct sunlight. If one isolates protozoa from the marine environment, one is very likely to get a flattened ciliate called *Euplotes*, belonging to a group called the hypotrichs, which move on compound ciliary organelles called cirri. Some species of *Euplotes* are cosmopolitan, occurring in both Northern and Southern Hemispheres. Addition of a few drops of baker's yeast suspension often results in denser cultures of marine *Euplotes*. Marine amoebae are often grown on agar made with seawater. In general, marine organisms may not tolerate such high temperatures as freshwater and soil organisms.

A simple classification

The Protozoa are no longer considered a single phylum but a group of phyla, whose classification has recently been revised (Levine *et al.*, 1980). A greatly abridged version of that detailed scheme is presented here, including only some groups made up largely of free-living species. The names in brackets are those of representative genera and do include a few major human pathogens. (The genus *Plasmodium*, to which malaria parasites belong, is classified in the phylum Apicomplexa, class Sporozoea, subclass Coccidia.)

Phylum SARCOMASTIGOPHORA
> With flagella or pseudopodia or both; with rare exceptions, a single type of nucleus in any one species; sexuality, when present, syngamy (fusion of cells).

Subphylum MASTIGOPHORA
> One or more flagella in trophozoites (active, feeding stage).

Class PHYTOMASTIGOPHOREA
> Typically with chloroplasts; if chloroplasts lacking, relationship to pigmented forms clear. (*Euglena, Peranema, Chilomonas, Chlamydomonas.*)

Class ZOOMASTIGOPHOREA
> Chloroplasts absent; no distinct and close relationship to pigmented forms. (*Bodo, Trypanosoma, Leishmania, Giardia.*)

Subphylum SARCODINA (some classes omitted)
> Pseudopodia or locomotive protoplasmic flow without distinct pseudopodia; flagella, if present, usually only in temporary stages.

Superclass RHIZOPODA
> Lobopodia, filopodia, or reticulopodia, or protoplasmic flow without distinct pseudopodia.

Class LOBOSEA
> Locomotive pseudopodia lobose, sometimes with filiform sub-pseudopodia; usually uninucleate; multinucleate forms not flattened or much-branched plasmodia; no spores produced; with or without tests (shell-like enclosure). (*Amoeba, Entamoeba, Saccamoeba, Acanthamoeba, Naegleria, Arcella, Difflugia.*)

Class FILOSEA
> Pseudopodia filose; with or without tests. (*Vampyrella, Euglypha.*)

Class GRANULORETICULOSEA

Delicate, finely granular or hyaline reticulopodia. Most species are Foraminiferida, marine organisms with tests.

Superclass ACTINOPODA

Axopodia with microtubular stereoplasm; often spherical; usually planktonic.

Class ACANTHAREA
Class POLYCYSTINEA
Class PHAEODAREA

Marine; usually with skeleton; inner region enclosed in capsular membrane. Polycystinea and Phaeodarea are the 'radiolarians'.

Class HELIOZOEA

Without central capsule; with or without skeleton of silica or organic material. (*Actinophrys, Actinosphaerium, Acanthocystis.*)

Phylum CILIOPHORA

Cilia in at least one stage of life cycle; two types of nuclei (macronucleus and micronucleus); sexuality involving conjugation or related processes.

Class KINETOFRAGMINOPHOREA

Oral infraciliature (pattern of bases of cilia) only slightly distinct from somatic infraciliature; cytostome (mouth) on surface of body or at bottom of atrium or vestibulum; compound ciliature typically absent. (*Coleps, Didinium, Colpoda, Nassula, Discophrya.*)

Class OLIGOHYMENOPHOREA

Oral apparatus, at least partly in buccal cavity, generally well defined; oral ciliature distinct from somatic ciliature and consisting of paraoral membrane and small number of compound organelles. (*Colpidium, Tetrahymena, Paramecium, Vorticella.*)

Class POLYHYMENOPHOREA

Conspicuous zone of organelles in oral region; cytostome at bottom of buccal cavity or infundibulum; somatic ciliature complete or reduced, or organised into cirri (brush-like structures). (*Stentor, Halteria, Euplotes.*)

References

1. Books
These are meant to be useful both to readers who want basic introductions and to those who wish to go more deeply. Some are no longer in print but may be available through libraries. A few books especially recommended to secondary school teachers are indicated by an asterisk, but other titles will also be useful to teachers with special interests, while the starred books may also be of value to advanced workers.

Bick, H. 1972. *Ciliated protozoa: an illustrated guide to the species used as biological indicators in freshwater biology.* Geneva: World Health Organisation.

Bovee, E. C. & Sawyer, T. K. 1979. *Marine flora and fauna of the northeastern United States. Protozoa: amoebae.* Washington: National Oceanic and Atmospheric Administration. (a key that will be useful in other parts of the world.)

Chen, T. T. ed. 1967-72. *Research in protozoology.* **1, 2, 3,** and **4.** Oxford: Pergamon.

Corliss, J. O. 1979. *The ciliated protozoa.* 2nd edition. Oxford: Pergamon. (Taxonomy and guide to the literature. Up-to-date. Not a key for identification.)

Curds, C. R. 1969. *An illustrated key to the British freshwater ciliated protozoa commonly found in activated sludge.* London: Her Majesty's Stationery Office.

Elliott, A. M., ed. 1973. *The biology of Tetrahymena.* Stroudsburg, Pennsylvania: Dowden, Hutchinson and Ross. (Essential for all aspects of this much studied ciliate.)

Fulton, C. 1970. Amebo-flagellates as research partners: The laboratory biology of *Naegleria* and *Tetramitus.* In: *Methods in cell physiology,* **4,** edited by D. M. Prescott, 341-476. London: Academic Press.

Giese, A. C. 1973. *Blepharisma.* Stanford, California: Stanford University Press.

Grassé, P. P. ed. 1952-53. *Traité de zoologie.* Paris: Masson. (Survey of protozoa other than ciliates.)

Grell, K. G. 1973. *Protozoology.* Berlin: Springer-Verlag. (English translation of German original.)

Griffin, J. L. 1978. Pathogenic free-living amoebae. In: *Parasitic protozoa,* **2,** edited by J. P. Kreier, 507-549. London: Academic Press.

*Jahn, T. L., Bovee, E. C. & Jahn, F. F. 1979. *How to know the protozoa.* 2nd edition. Dubuque, Iowa: Wm. C. Brown.

Jeon, K. W., ed. 1973. *The biology of amoeba.* London: Academic Press. (Descriptive and experimental reviews, mostly dealing with larger amoebae such as *Amoeba proteus.*)

Jones, A. R. 1974. *The ciliates.* London: Hutchinson.

Kudo, R. R. 1966. *Protozoology.* 5th edition. Springfield, Illinois: Charles C. Thomas. (Out-of-date but much consulted for generic identifications.)

Leedale, G. F. 1967. *Euglenoid flagellates.* Englewood Cliffs, New Jersey: Prentice-Hall.

Levandowsky, M. & Hutner, S. H. eds. *Biochemistry and physiology of protozoa.* 1979-1981. **1, 2, 3,** and **4.** London: Academic Press (Advanced and valuable reviews.)

*Mackinnon, D. L. & Hawes, R. S. J. 1961. *An introduction to the study of protozoa.* Oxford: Oxford University Press. (A well-written and well-illustrated text which can still be consulted with profit. Includes sections on use of the microscope, culture, and staining, and suggestions for study of some of the organisms included in the present publication.)

Ogden, C. G. & Hedley, R. H. 1980. *An atlas of freshwater testate amoebae.* London: British Museum (Natural History) and Oxford University Press. (Scanning electron micrographs of many testaceans. Not a key, but informative and beautiful.)

Page, F. C. 1976. *An illustrated key to freshwater and soil amoebae.* Ambleside: Freshwater Biological Association. (Includes methods of culture and observation.)

*Parker, S. P., ed. 1981. *Taxonomy and classification of living organisms.* 2 vols. New York: McGraw-Hill. (Includes brief articles on all taxa of organisms including protozoa, down to the level of family, written by authorities.)

Patterson, D. J. 1978. *Kahl's keys to the ciliates.* Bristol: published by the author. (A translation of Kahl's keys down to genera. These keys are still much used in the original German. Useful if changes in ciliate taxonomy since 1935 are kept in mind. Available only from the author at Department of Zoology, University of Bristol; £2 or US$5.)

Rainer, H. 1967. *Urtiere, Protozoa, Wurzelfüssler, Rhizopoda, Sonnentierchen, Heliozoa: Systematik und Taxonomie.* Jena: Gustav Fischer Verlag. (Key to heliozoans, more complete than any in English.)

Schuster, F. L. 1979. Small amebas and ameboflagellates. In: *Biochemistry and physiology of protozoa,* 1, 2nd edition, edited by M. Levandowsky and S. H. Hutner, 215-285. London: Academic Press. (A valuable review of many aspects, dealing primarily with *Acanthamoeba, Naegleria,* and *Tetramitus.*)

*Sieburth, J. McN. 1979. *Sea microbes.* New York: Oxford University Press. (A survey of prokaryotes and eukaryote protists in sea water.)

*Sleigh, M. A. 1973. *The biology of protozoa.* London: Edward Arnold. (One of the best modern introductions, easily read through while valuable as a reference. Includes a survey of major groups but not intended as an aid to identification.)

Tartar, V. 1961. *The biology of Stentor.* Oxford: Pergamon. (The classic work on this splendid ciliate, by the authority who devoted many years to experimental investigations.)

Van Wagtendouk, W. J. ed. 1974. *Paramecium: a current survey.* Amsterdam: Elsevier. (Reviews of many aspects: taxonomic, morphological, experimental.)

*Vickerman, K. & Cox, F. E. G. 1967. *The protozoa.* London: John Murray. (Many students have been introduced to the protozoa through this slender, well-illustrated work, which deals with representatives of free-living and parasitic groups.)

2. Papers

Belcher, H. 1978. A miniature battery-operated centrifuge. *Microscopy,* **33,** 278-279.

Jones, A. R. 1980. Using the ciliate protozoan *Vorticella* in teaching. *Jnl Biol Educ.,* **14,** 119-126.

Levine, N. D., Corliss, J. O., Cox, F. E. G., Deroux, G., Grain, J., Honigberg, B. M., Leedale, G. F., Loeblich, A. R., III, Lom, J., Lynn, D., Merinfeld, E. G., Page, F. C., Polyansky, G., Sprague, V., Vavra, J., Wallace, F. G. & Weiser, J. 1980. A new revised classification of the protozoa. *J. Protozool.,* **27,** 37-58.

Nanney, D. L. & McCoy, J. W. 1976. Characterization of the species of the *Tetrahymena pyriformis* complex. *Trans. Amer. Micros. Soc.,* **95,** 664-682.

Index

Note: Some trivial mentions of subjects included in this index are *not* cited and *not* meant to be cited.